Enhanced Evolution

The Miracle of Using External Information

Steve Thompson

ISBN – 13: 978-1519556370
ISBN – 10: 1519556373

DEDICATION

To the millions of species and our countless ancestral life forms that participated in the 3.5 billion year, trial-and-error process of tiny, incremental advancements that evolved us to the point where we no longer need to wait for random, beneficial, genetic mutations to propel us forward; they endowed us not only with the capacity to choose alternative destinies, but to analyze the very process that created us.

Contents

Information Is a Special Kind of Resource

I first started thinking about the nature of information back in 1971 when I was an editor for Instruments & Control Systems Magazine, which was directed at process control engineers. Process control is the sector of industry that deals with bulk processing of products, such as petroleum, paper, food, etc. The other sector of the manufacturing industry is the Original equipment market, or OEM. It is characterized by the production of discrete units, such as screws, automobiles, refrigerators or computers. While trying to understand how all phases of process control relate to each other, I developed a model that described the structure of the entire manufacturing industry, including process control. In a February 18, 1971 memo, I created a diagram that illustrated mankind's production from natural resources to finished goods and called it Resource Utilization.

It depicted how man gathers (mines, harvests, herds, etc.) earth's resources, processes (refines, purifies, assembles, machines, etc.) them into finished goods, packages them to preserve their utility during distribution, and ships them to end users. As an outgrowth of that analysis, I postulated that there were three "special cases" of resources: people, energy and information. The common denominator that makes them special is that they are required at virtually every step of any manufacturing process, no matter what the end product.

6

Nothing happens without people planning, making decisions, operating machines or overseeing processes in an ever-changing environment. Energy is required at every stage, be it extracting raw materials, refining them, assembling them, or just shipping stuff around. Information is also required everywhere. It is needed to control the instruments and machines and to inform the people involved at every step. The knowledge of how a process works comprises the information that is used to guide the entire operation of every industrial plant.

I came to realize that information was its own special case, among special cases of resources. I postulated that as a resource, information has three unique characteristics. It can be purely abstract; it can be man made; and it is cumulative. This does not imply that information is not a "natural" resource. It is generated by nature and by at least one of nature's creatures, humans; therefore, it is perfectly natural, but not necessarily physical. It can also be abstract, which has very far-reaching implications. We also don't seem to be able to get out of the way of accumulating more and more of it.

I concluded that human "progress" is dependent on the ability to build upon and disseminate prior knowledge, so that each generation does not have to reinvent the wheel; it can continuously improve upon the wheels it has already developed. I suggested that perhaps the reason that man progresses faster than the other animals is because he is the only one capable of any significant resource generation, i.e., information libraries. He does not have to wait for changes in genetic information transfer (evolution) to cope with his environment; he can chose to alter his environment, or enhance his capabilities, instead. Though I felt that information and evolution were somehow connected, I did not think any more about it at the time, so that idea lay dormant for over 40 years.

Fast forward to the summers of 2013 and 2014 when I began developing courses on the Big Bang and Inventions for the Baldwin-Wallace University Institute for Learning in Retirement. That exercise presented an opportunity to reflect upon and reconsider information's characteristics and importance in an evolutionary sense.

The Life-Information-Evolution Connection

How are we defining information? It is a lot like money, it's hard to define, but everyone knows it when they see it. For our purposes I am using an everyday definition distilled from multiple sources. It includes both genetic and non-genetic examples of information, which both turn out to be important:

Basically, information is a stimulus that has meaning in some context for its receiver. Information is facts learned about something, such as data, statistics, instruction, news, knowledge or intelligence. It includes what is conveyed or represented by a particular arrangement or sequence of things, such as genetically transmitted information. Similarly, when information is entered into and stored in a computer, it is generally referred to as data. After processing, output data can again be perceived as information.

Again, information is a stimulus that has meaning in some context for its receiver. It also implies that information allows the holder of that information to make predictions, with an accuracy better than chance.

What does this have to do with living things? In his book, *The Evolution of Everything*, while discussing Watson and Crick's great work on DNA, Matt Ridley expresses the connection between information and living things eloquently. "Vitalism'...is the old idea that there is something peculiar and special about living tissue...some mysterious vital ingredient that makes it 'alive.'...In a way you could argue that the double helix did confirm that there is something peculiar and special about living tissue – namely, that it contains digital information capable of both replicating itself and instructing the synthesis of machinery for harnessing energy. The secret of life, unexpectedly, turned out to be an infinitely combinatorial message written in digital form in three-letter words in a four-letter alphabet. This was very much not what vitalists had expected; it seemed too mundane – though actually it is one of the most beautiful ideas every to cross a human mind – that **life is information.**" (My emphasis)

We must be careful about the order of terms. Though all life is information, all information is not life. Life is a subset of information. A way to specify that subset is to stipulate that **life is information that can copy itself.**

Now, if we accept that:

Life is Information that can copy itself,

then it follows that:

Evolution is *Changes* in that Information.

Evolution can be initially defined as changes in a DNA's information package. No changes; no evolution. That is Stage 1 Evolution. Those changes can involve modifications of existing information or the addition of new information. All changes are not guaranteed to be beneficial; the vast majority is not. However, as the more successful implementations thrive and the less successful are extinguished, evolution's path inevitably leads toward developing ever more successful and sophisticated life forms.

However, there are two sources of information in play that both affect life forms: a) internal, biological, DNA information and b) all other information that is external to DNA. The premise of this paper is that *an organism that acquires the ability to utilize information external to its DNA enters*

Enhanced (Stage 2) Evolution

Stage 1 evolution can be considered purely biological, DNA dependent, entry-level evolution. An organism that learns to process and utilize information in addition to that contained within its own DNA greatly enhances its chance for success in a hostile universe. It enters Stage 2 and elevates its game by acquiring additional, non-DNA-limited, competitive advantages. They include the ability to alter its environment to be more in its favor, which is really just another competitive advantage.

The amount of external information available for an organism to utilize is virtually unlimited, compared to what its DNA contains. In the enhanced evolution stage, evolution is also driven by changes in that information; changes in the amount and quality of the information the organism can acquire, store, disseminate, access and interpret to its advantage.

For our purposes, we will designate information contained within, or accessible to, DNA as "internal" information; all other information outside of DNA, whether it resides inside or outside of the life form, will be considered "external" information.

In *Why Information Grows*, Cesar Hidalgo gives a beautiful illustration of the difference between internal and external information. "Consider two types of apples: those you grow on trees and you buy at the supermarket, and those that are designed in Silicon Valley. Both are traded in the economy, and both embody information, whether in biological cells or silicon chips. The main difference between them is not their number of parts or their ability to perform functions – edible apples are the result of tens of thousands of genes that perform sophisticated biochemical functions. The main difference between apples and Apples is that the apples we eat existed first in the world and then in our heads, while the Apples we use to check our email existed first in someone's head and then in the world. Both of these apples are products and embody information, but only one of them – the silicon apple – is a crystal of imagination."

The Apple is just one of our inventions, our crystals of the imagination, that surely enhance our evolutionary chances. The sum total of these crystals that originate in our heads is staggering, and their effects are profound.

Understanding Information's Role

Cosmology teaches us that we are a very fortunate life form. We have evolved to live in our own "Goldilocks Zone" where everything is "just right" for our survival on the surface of this particular planet. Life was precariously established over 3.5 billion years ago along with some primordial DNA. Humans evolved late in the game and though we may currently be at the top of the food chain, no earthly species is destined to persevere forever, if for no other reason than that earth won't last forever. All kinds of unavoidable, catastrophic, events, shown in Table I, will impact our specie's future and can terminate our existence long before earth disappears.

If I had to bet, I'd wager that the chances we will annihilate ourselves, along with a lot of other life forms, are greater than the chances that we will survive to experience one of the other inexorable, catastrophic events. Assuming we turn out to not be our own worst enemies, glaciation cycles will probably prove to be more of an inconvenience than species threatening. Asteroid/meteor impacts are unpredictable. Those of species threatening consequence are predicted to occur, on average, every 45 million years. We can probably develop detection and deflection systems to thwart that danger if we put our minds to it. Supercontinent formation may not be species threatening but will certainly stress us with climate changes and tectonic upheaval.

A nearby supernova could produce enough gamma radiation to deplete our ozone layer and render earth's life

TABLE 1
CATASTROPHE TIMESCALE

Cause	Time Frame (years)
Human Induced disaster	?
Glaciation (10 deg/cycle)	every 100,000
Asteroid/Meteor Impact	45 million
Supercontinent formation	50-350 million
Supernova	100 million
End of Photosynthesis	0.6 - 1.3 billion
Loss of Oceans	1.1 billion
Large Axial Tilt	1.5 – 4 billion
Loss of Magnetosphere	3 – 4 billion
Red Giant Phase of Sun	4.5 billion

forms defenseless against our sun's ultra violet radiation. It would be a beautiful, but deadly sight. A sure-fire bet for species annihilation is that as the sun enlarges it will effectively move our Goldilocks Zone out beyond earth's orbit as it continually makes earth hotter and hotter. Between 0.6 – 1.3 billion years all plant life will disappear, taking all animal life with it, and our oceans will evaporate. Nature's backup plan is that any of the other catastrophic events will wipe us out if we last that long.

How might we escape, or at least postpone, species annihilation? We are in a box. Our Goldilocks Haven is slated for the dustbin of galactic history and we can't evolve our way out of eventual major asteroid hits, the end of photosynthesis or the sun's increasing heat load. Nature cannot modify us fast enough for us to acquire characteristics that would enable us survive the vastly different situations and environments that will surely, and sometimes suddenly, present themselves. What hope do we have? If not forever, how long can we postpone our extinction?

No matter how you slice it, over the very long term there is only one real option for prolonging our specie's survival – move on to other Goldilocks Zone look-alikes. To be optimistic about extending our long-term survival chances we cannot rely on DNA driven evolution; we will need a better strategy. We may already have just the special capabilities we need to draw upon.

The chronological sequence of inventions in Table 2 is a record of human information generation, storage, dissemination and access, which ultimately enables us to

TABLE 2 – Sequence of Inventions*

	Year	Gener- ation	Storage	Dissem- ination	Acces
Consciousness	?	X			
Memory	?		X		X
Speech	1.75 b yr			X	
Language	200k BC	X		X	X
Written Language	3100 BC	X	X	X	X
Libraries	2600 BC		X	X	X
Paper	105 CE		X	X	X
Printing Press	1282			X	
Telegraph	1838			X	X
Photography	1840	X	X	X	X
Telephone	1876			X	X
Radio	1895	X		X	
Television	1927	X		X	
Photocopier	1938		X	X	
Digital Comm.					
Transistor	1947				
Computer	1955	X	X	X	X
Cell Phone	1973	X	X	X	X
Internet	1980s	X	X	X	X
E-mail	1993	X	X	X	X

* Many of these inventions are described in *100 Greatest Science Inventions of All Time* by Kendall Haven, which I used as a text for a course on Inventions and which was very helpful in compiling this list and the descriptions of the inventions.

use that information to enhance our survival chances. At each step our ability to generate, store and access (retrieve) a growing volume of accumulated information increases. If evolution is changes in information, either as an alteration to existing information, or as acquisition of new information, adding information to the external information portion of our total known information reservoir is where the largest amount of change can happen, most rapidly. It is where we can, and have, effected the most change in recent human history.

As Hidalgo points out, our products are not just the physical embodiment of information, they are, "the vehicles we use to communicate something more important than messages: the practical uses of knowledge and knowhow." The specific inventions I chose excel at that task. Hidalgo differentiates between knowledge and knowhow, "Simply put, knowledge involves relationships or linkages between entities...knowhow is different from knowledge because it involves the capacity to perform actions." Our inventions are where the rubber hits the road; it is where knowledge and knowhow come together to create never-before-seen products.

Our inventions are the end products, of using and manipulating external information. They propel us forward in our capabilities to thrive; they underpin our cultures. Let's see how we have progressed in our abilities to manipulate external information to invent our way to success. It is quite an impressive ride.

Table 2 shows the dates of the inventions and the areas where they made their greatest contributions to generating, storing, disseminating and accessing information. Consciousness and memory are both biological features that humans are all born with. Mother Nature invented them; we, one of nature's offspring, invented everything else on the list. Significant inventions become more frequent over time, because our base of stored information continually expands. That enables us to generate ever more sophisticated, more useful inventions. What has this particular, information-related progression of inventions done for us?

Consciousness is the source of all human, information generation. In combination with our memory, it produced all of the other inventions on the list. It is the state of awareness that permits one to be aware of external objects or something within oneself. It takes us beyond simple, instinctive responses and enables us to analyze and respond to external stimuli in a deliberative, non-instinctive way. It enables processing information to predict outcomes. It is the gateway to everything else.

Memory probably developed almost simultaneously with consciousness. Memory is the process by which we encode and store information, then locate it and retrieve it later. That keeps us from constantly having to reinvent the wheel. Without it we could not learn, store knowledge or make predictions with accuracy better than random chance. It's absence would short-circuit consciousness' potential to enhance our evolutionary prospects.

Speech arose about 1.75 billion years ago in ancestral species. Early speech was undoubtedly sign language, probably augmented by grunts, facial expressions and gestures. Oral speech enables communication, or sharing of information, between individuals or groups within sight or earshot.

Language developed between 150,000 – 300,000 years ago. It is a real game-changer. We could communicate and share complex ideas among groups and coordinate actions to defend ourselves or to hunt others. It is the basis for all human communication going forward.

Written language emerged about 3100 BC in Mesopotamia. It is a second game-changer, because information became permanent and could be sent over long distances. Written information greatly expands our data base storage capacity from just limited, inaccurate, individual memories to everything learned by everyone, which can then be passed along to others.

Libraries aggregate and archive written information for indefinite periods. This stored information can contain vast data about an entire culture and be accessed at any time by many individuals. It is centrally accessible, permanent, massive memory.

The Printing Press is arguably the most important invention of all time. This was the first method for disseminating large quantities of information to anyone. Three hundred copies of Gutenberg's original 1,282-page Bible took three years to produce. The only other method of reproduction, hand copying, would have taken 400 times longer. It also made mass literacy and education both possible, and necessary.

The Telegraph was the first timely, long distance communication. We could send coded messages at the speed of light, any time, between any two points where we could string a wire.

Photography freezes information for all time in all its detail. It replaces lengthy verbal descriptions of objects or events with an instant, more accurate portrayal. If a picture is worth a thousand words, think of what a moving picture, a sequence of still pictures, is worth.

The telephone one-upped the telegraph by giving us widely disbursed, instantaneous, person-to-person, voice communication, replacing encrypted Morse code.

Radio enabled instant mass communication anyplace, without wires, over any distance. It changed communications between countries and spread news and entertainment in new ways that changed our culture.

Television went radio one better and added moving images. Message content and dissemination exploded. It is like having eyes and ears instead of just ears.

Photocopying creates a virtually infinite supply of exact copies of an original document. It is indispensible to modern life. We make hundreds of millions of copies of something every day.

Digital Communication, spawned by the invention of the transistor, includes any combination of people and machine communication: computers, e-mail, cell phones, the Internet, you name it. Their storage capacity is beyond anything traditional libraries can contain. The accessibility of information is unmatched; one can scan billions of sources in less than a second to search for virtually any scrap of information. It puts our ability to search and manipulate information on steroids.

The Information/Evolution Relationship

How are evolution and human inventions interrelated? The net, net, net of all of this is that two distinct classes of information have evolved. What matters in this discussion is where information **resides**. One reservoir is **biological**, or **internal** to a living organism's DNA. The other is any information **external** to a life form's DNA. It can be said that **changes** in information **are** the total story of evolution; what evolution is all about. But there are two reservoirs of available information that are undergoing those changes.

The earliest, simplest life forms begin with only biological, DNA information. They are capable of only reflexive behavior. As evolution plays out, they develop instinctive behavior, as changes to DNA that provide beneficial adaptations are passed on by the more successful survivors and those that are less beneficial are phased out with the less adept. All reflexive and instinctive behavior is based upon innate biological factors, without being based upon prior experience (that is, in the absence of learning). Adding internal information to DNA is a glacial process, and not within a specie's control. It takes eons to "build" a more complex, more successful DNA.

When consciousness arises, the game changes. Consciousness, in combination with memory, enables learned behavior. It enables storage of any external information that its mind can decode that is meaningful to it. When consciousness begins to recall data, analyze it,

20

and make choices that are better than random chance, that is the genesis of Enhanced, or Stage 2, Evolution. As consciousness develops, the reservoir of external information a life form accumulates becomes more and more important on a day-to-day basis, because it permits anticipation of, and planning for, the future, i.e., determining more favorable outcomes.

At that point, it is clear that if evolution is changes in information, the greater opportunity for rapid evolutionary progress is to continually add to our store of external information. It represents an unlimited source of additional information, found throughout the universe, that a conscious mind can avail itself of to use to its advantage as rapidly as it can decode it. It dwarfs the amount of information DNA can ever accumulate, and the rate at which it can be acquired and disseminated.

The entity begins to free itself from the constraints of its DNA and use external information to provide better chances for its survival. Its conscious mind is the information processor that provides the bridge of accessibility between the entity and all external information. The entity no longer needs to wait for random, beneficial, genetic mutations to propel itself forward; it attains the capacity to choose alternative destinies. With enough development, it might access its own internal information, its very own DNA.

External information can be uncovered by discovery or by reasoning. Once generated, it can be stored and analyzed for its usefulness. To maximize its usage by the species it has to be disseminated. Finally, it has to be recalled and applied – accessed and used - in some way, or it is just another dead end, a brilliant equation written in the sand. Our inventions are evidence of our information being applied for use. They are the products of our learning from our information and implementing our knowledge in never-before-seen ways.

Consciousness and memory are emergent properties of our brains. Everything else on our list of inventions is man-made, using information that originates externally to our bodies. The gift of the combination of consciousness plus memory is the ability to recognize and use external information. We owe our position at the top of the evolutionary chain to both internal and external forms of information. However, each represents a different stage of evolution, and the sum of the two is greater than its parts.

Let's look at the differences between the characteristics of internal and external information. All earthly life forms are DNA based. DNA powered evolution is a slow, strictly trial and error process, but it works. We have gone from the most primitive, DNA-driven, life forms, such as cyanobacteria, to humans in about 3.5 billion years. That's very impressive; it is the consequence of myriad mutations of some primordial form of DNA.

DNA, illustrated in Fig. 1, is basically chemically encoded information. It is an instruction manual, for building a life form. Nature has created a molecular library of genetic information in each of us – a genome that is our entire biological, information bank, our life. DNA's storage capabilities and adaptability have evolved by trial and error over billions of years until it got large enough and complex enough to formulate a recipe for humans.

It is a marvelous instruction manual. Its language is composed of only four "letters," the four nucleotides: adenine (A), cytosine (C), guanine (G) and thymine (T). G and C are always found in combination, as are A and T. Those letters only combine into four possible "words," GC, CG, AT and TA. That is a fairly limited language. Nonetheless, it has achieved remarkable complexity.

Figure 1 – DNA Construction

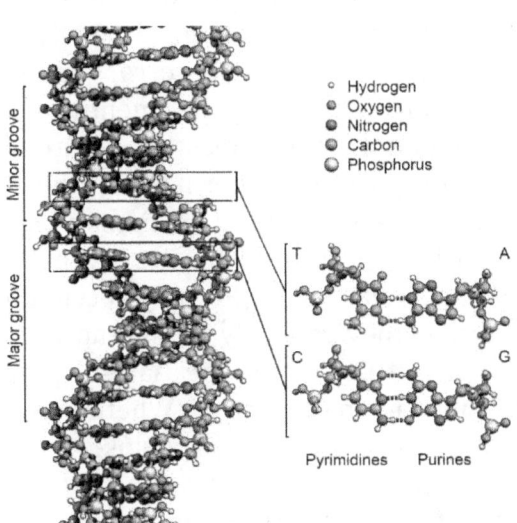

Fig. 1 Structure of the DNA double helix. The atoms in the structure are color-coded by element and the detailed structures of two base pairs are shown in the bottom right. From Wikipedia.

Wikipedia suggests the following frame of reference:
An analogy to the human genome stored on DNA is that of instructions stored in a book:
- The book (genome) would contain 23 chapters (chromosomes);
- Each chapter contains 48 to 250 million letters (A, C, G, T) without spaces;
- Hence, the book contains over 3.2 billion letters, or 1.6 billion words, total;
- The book fits into a cell nucleus the size of a pinpoint;
- At least one copy of the book (all 23 chapters) is

contained in most cells of our body. The only exception in humans is found in mature red blood cells, which become enucleated during development and therefore lacks a genome.

DNA is nothing if not persistent. It keeps recombining its four words by trial and error in semi-infinite permutations, unconsciously experimenting to create the next best adaptation to the existing environment. After about 3.5 billion years of tinkering with the system, it is now sophisticated enough to build human beings. That is quite a remarkable accomplishment given the language limitations, and that changes only occur by random chance, and that they can only be spread by inheritance. Right now, we think that the latest, greatest, DNA adaptation to living on earth is us. Whether we are only at the beginning of our reign, or close to terminating ourselves, remains to be seen.

We saw that information can be physical such as DNA, or abstract such as the theory of relativity. It can also be created, unconsciously, in nature, such as magnetic fields that guide the flights of certain birds, or consciously, by man, such as street signs. Where evolution is concerned, the two classes of information we are dealing with are biological information that is generated, stored and utilized entirely *internal* to DNA, and all other information that is created, stored and used, not just within a mind, but *external* to living organisms.

I separate internal and external information into separate categories, if for no other reason than that DNA cannot store or manipulate man-made information. All man-made information is the product of a conscious mind, generated external to DNA. A large portion of external information is abstract. Since abstract things are not physical, they cannot be contained within DNA. Though

DNA cannot manipulate man-made information, the converse is not true; we are beginning to decode and manipulate DNA, itself. Again, the key is that DNA information is internal to the organism, while man-made information is external to its DNA.

Since we are only dealing with information as it applies to evolution I will use "internal" as a synonym for "DNA," information. Since we are only dealing with human beings, I will use "man-made" as a synonym for any "external" information. Though many other creatures use external information with varying degrees of sophistication, and though there is undoubtedly much external information we have not yet decoded, we are only considering the ever-expanding body of information already available for us to use at any given moment.

One could argue that since DNA produced our minds, all information, even our sensory interpretations of the "real" world and our abstract thoughts are products of DNA, but that won't make any difference in the overall importance of external information to evolution; it's still external to DNA, itself. We will take up the importance of abstract information later.

Comparing Internal and External Information

The significance of internal vs. external information is profound. DNA is an internal, living, biological, information bank with strict limitations. Again, evolutionary changes are limited to painstaking, random adjustments to any life form's DNA, information bank.

Even as the most highly evolved species on earth, about 95% of our brains are devoted to subconscious body maintenance, of which we are completely unaware. We do not think about or consciously control our blood circulation, breathing, growth, muscle manipulation, cell reproduction, digestion, etc. – almost everything we do is done for us by our subconscious, completely beyond our control. However, we have somehow evolved to a level of cognition where a part of our brain becomes our "mind." In our minds, we can process and/or create something that DNA cannot - external information – information that can be stored, retrieved and exchanged *outside* of not just DNA but outside of the organism, itself. That is an evolutionary game changer! Table 3 compares the characteristics of the two information systems, using "DNA" and "Man Made" as stand-ins for "internal" and "external."

Base Language

Both DNA and man-made information tend to be cumulative. A major difference is their robustness. Our basic language contains a minimum of 42 characters: 26 letters, plus Hidalgo's additional six punctuation characters, plus 10 numbers. DNA only combines four

TABLE 3 - Comparing Information Systems

	DNA	Man Made
Base Language	4 letters	>42 characters
Accumulation Rate	Random Mutation	Exponential
Duration	Life of Carrier(s)	Indefinite
Replication rate	Internal: World Class External: Procreation	Unlimited
Dissemination Rate	Procreation	Speed of Light

letters into four words; by comparison, our language is semi-infinite. Like a computer language that would use 42-bit words instead of 4-bit words, it has the potential to manipulate and transmit much more complex information, much faster.

Accumulation Rate

The amount of biological information stored in a genome is relatively stagnant, and can only be varied or increased via widely spaced, random mutations. External information, manipulated in conscious minds, is rapidly cumulative. 1 + 1 = 2 led to 3 x 6 = 18 and E = mc2, etc., at light speed compared with DNA evolution. Forty-two characters led to War and Peace, the constitution, and relativity; 8 musical notes lead to Ring Around the Rosie, In The Mood and Beethoven's 5th, all in less than 50,000 years. The growth of our external information bank is exponential; it does not have to wait eons for random mutations that may, or may not, improve the survival chances of its carrier.

Duration

Any DNA information library perishes with its host. It must also be recreated in its entirety with each generation. Barring burning every book and cutting every communication line, human information lasts indefinitely.

Once we learn how to add 2+2 and write it down for all to see, nobody ever has to figure it out again. Everyone can use it and build upon it without ever recreating that wheel. Once we print a million copies of a book or put something on the Internet, you almost can't get rid of it, as Facebook continually demonstrates to the chagrin of many.

Replication Rate

Internally, DNA is amazing. It can make trillions of copies of itself, year-after-year, virtually error-free. The math is stunning: A baby is born with about 2 trillion cells, all originating from, and manufactured under the control of, its original, single cell's DNA at conception. During nine months in the womb, not considering cell replacements, the baby's DNA averages a production rate of 80,000 cells per second, every second. That's prodigious!

It is even more amazing. While DNA is mass-producing cells, it is also delineating them into over 200 distinct types of cells, each with a different mission. It produces the proper proportions of each type of cell and assembles them into a coordinated, functioning, living entity. It is a monumental tour de force. And that's only the beginning! An adult with 50-70 trillion cells needs to produce two million red blood cells per second just to keep up with red blood cell depletion. Though blood cells don't contain DNA, they are produced in our bone marrow using DNA's instructions. Overall, the body is cranking out about 3 million new cells per second, every second, one-third of which carry their life form's, fully reproduced, DNA library. There are about 2.5 billion seconds in an 80-year-old's life. That translates into about 6×10^{21} (six sextillion) cell reproductions, a third of which carry an exact DNA copy. Clearly, DNA is a world-class duplicator of itself, within a given life form. It makes Six Sigma reliability look like child's play.

However, it can only generate a copy of a complete life form via procreation. Even then it is not a carbon copy but a composite of two parental DNAs. Einstein can't duplicate himself, or Mozart. Fortunately, neither can Jack The Ripper or Hitler. Our external communication can generate virtually unlimited, exact duplicates of any information and distribute them to everyone, almost instantly.

Dissemination Speed

A DNA, chance improvement (or detriment) - a species-beneficial piece of information - can only be transmitted and spread by inheritance, via procreation, to direct descendants. To spread to every living human being, it would take a minimum of 25 generations to reach over seven billion of us. That assumes that everyone marries, averages 2.5 children, and those children only marry people who have not yet inherited the mutation. At the current estimate of 29 years per generation, according to Ancestry.com, that would take a minimum of 725 years, and that is a grossly unrealistic, best case, scenario.

The ability to disseminate external information makes a critical difference. Prior to the development of language, human progress was limited to monkey see-monkey do. Our brains were our only libraries and they are not the most reliable or extensive. Though that information transfer is still a lot quicker than procreation, we could only communicate a behavior or a skill by copying others. Language greatly sped up communication. Writing added permanency, and provided archival capability of information for anyone's use, on demand. One mind's capacity is no match for our external storage capacity, permanency and accessibility. The path from the printing press to electronic data transmission made information transfer universal and instantaneous.

29

Externally, what any one person can add to humanity's existing data bank can be disseminated to everyone, simultaneously, at the speed of light for instant use. We completely bypass the generations it takes for DNA to pass along a new development to every human being.

Volume – The 600-Pound Gorilla

You can only pack so much information into a given number of atoms. I don't know what the information/atom correlation is (one can postulate that an upper limit is one bit per atom), but it seems certain that more atoms have the potential to hold more information than fewer atoms. Since DNA is restricted in its size, and our external sources of information include all the matter and energy in the universe, there is a whole lot more information potential outside our bodies than inside. As we learn to decode and use large amounts of that potential information, we surely begin to arrive at a situation where we "know" more than DNA can know. With know-how, we can even think about building our own DNA. We are accumulating knowledge, and know-how, that our relatively ignorant ancestors, could only ascribe to the province of imaginary, all-knowing deities. How has accessing that virtually unlimited source of information played out so far?

How much information can a genome contain? Estimates of the number of atoms in DNA center around 204 billion, but word content is probably a more realistic measure. The printed version of the Encyclopedia Britannica contains about 44 million words; its online edition totals about 55 million words. Our DNA contains about 1.6 billion words (combinations of A, C, G & T), so Britannica would have to expand to 30 times its present size to rival the word content of a single DNA.

By contrast, Wikipedia contains over 4 million articles containing 2.5 billion words. I don't know how "efficient"

Wiki is in its use of words as information compared to DNA, but things are getting interesting. We have created at least one external databank that compares, in size, to DNA, but we have many others to consider in Table 4.

TABLE 4 (Bytes Source: Wikipedia)
HUMAN INFORMATION DATABANKS

Print Collections of Library of Congress
10^{12} bytes   = 1×10^{11} words   =   62.5 DNAs
National Climactic Data Center
4×10^{14} bytes   = 4×10^{13} words   =   25,000 DNAs
US Academic Research Libraries
2×10^{15} bytes   = 2×10^{14} words   =   125,000 DNAs
Hard disk capacity created in 1955
2×10^{16} bytes   = 2×10^{15} words   = 1.25×10^{6} DNAs
All printed material in the world
2×10^{17} bytes   = 2×10^{16} words   = 1.25×10^{7} DNAs
Total information volume generated in 1999
2×10^{18} bytes   = 2×10^{17} words   = 1.25×10^{8} DNAs
All words ever spoken by humans
5×10^{18} bytes   = 5×10^{17} words   = 3.12×10^{8} DNAs

Table 4 assumes that it takes two bytes per character to encode a word and that a word equals five characters. It then equates the words to DNA words, using 1.6 billion words/DNA. The power of using external information is that there is so much of it, compared to internal information. Even if only a fraction of it is "useful," in that it is being applied to enhancing our survival chances by adding to our capabilities to compete, it is comparatively unlimited. Using enough of it will eventually overwhelm what DNA can do. This becomes more obvious when we consider that we have accumulated enough information that we can already decode and begin to alter DNA, itself.

Table 4 shows that our Library of Congress, alone, contains the word equivalent of 62.5 DNAs. US Research Laboratories contain the equivalent of 125 thousand DNAs. The information that humans generated in 1999 equaled 125 million DNAs. Again, if only one percent of it was applied to better transportation, food management, weather forecasting, disease control, improved shelter and living conditions, etc., that is the equivalent of what a million DNAs couldn't do! And it is cumulative. And the gap between what DNA can contain and what humans can access externally will only widen.

Abstract Information – A Hidden Ace

We have another external information ace in the hole, abstract information. Abstractions are not concrete, physical things. They exist only in the mind. Only the mind can create abstract concepts such as justice, poverty, mathematics, music or speed, which information cannot be found in DNA or any physical information. Physical things behave according to rules that they cannot violate. Only the mind can speculate about unseen possibilities. Only minds can create arbitrary rules of conduct, government, or culture. That also means that only something with a mind can make an error, or on the bright side, avoid an error, i.e., make a better or a worse prediction or choice.

Concrete, real, information has limits. It is constrained to physical forms. Abstract information is unlimited. Though abstract thought is a product of our mind, within our brain, what those thoughts represent exist external to, or independent of, any physical form.

Our use of abstract information starts with our sensory interpretations of our world. Our brains interpret different wavelengths of light as colors and different wavelengths of mechanical vibrations as sounds. Those are abstractions we use to help us perceive our surroundings; they are not "reality;" they are interpretations; mind-generated perceptions that represent physical reality to better enable us to survive.

All abstract information that humans possess is man-made. Our minds generated every bit of it. One school of thought contends that all abstract information – any thought we will ever have - already exists, whether or not we ever uncover it. By that reasoning, what we know of the abstract is not "man-made," it is "man-discovered." Fortunately, for our purposes it doesn't make any difference, either way. Whether we generate it ourselves or merely discover something that pre-existed, any abstract information humans possess is always external. We acquired it via our mental processes and it is now ours to use. Anything abstract that we have not generated or discovered lies outside our human external information base and is not yet material to our evolution, anyhow. In our discussion, we can consider any external information we have, either real or abstract, as man-made.

To illustrate just how powerful the human capability for abstract thinking is, let's begin with some data from Hidalgo's article, *How Much Information Can The Earth Hold,* in the August 2015 issue of Scientific American. Think of the universe as a computer, as Seth Lloyd did in 2002. He concluded that the largest possible computer would be the universe. If every atom of our universe could contain one bit of information, it could hold a whopping 10^{90} bits of information. That's a trillion times itself seven times, times another million. Earth could contain about 10^{44} bits, which is a miniscule fraction of the universe's capacity.

In 2011 Hilbert and Lopez published an estimate of the cultural information stored in our planet's texts, pictures and videos. They concluded that as of 2007, humans had stored 2 x 10^{21} bits, or two trillion gigabits (100 billion DNAs), trivial compared to the earth's theoretical capacity.

Now consider that we can contemplate something abstract like the set of whole numbers (or we couldn't contemplate quantities like 2 X 10^{21} bits). How many whole numbers are there? To answer that, we have to generate another abstract thought, infinity. Because we can abstractly reason that we can always add one to any number, we can deduce that no matter how large any number gets, we can always generate another, larger number, endlessly. The counting process becomes what we call infinite; we cannot comprehend a largest number, because for all practical purposes, we can't ever get to a last one. One can argue that nature doesn't "count" at all. It doesn't care how many "things" there are. Humans invented counting numbers to make sense of things on their terms, in their reality.

Consequently, if you tried to store all of the whole numbers in the real number system within the largest computer in the universe, that single task would overwhelm the universe's storage capacity all by itself! It would run out of storage long before it ran out of numbers to store. We can easily imagine an infinite number of things, any one of which easily exceeds the capacity of the universe to store it. Any irrational number such as pi, which never repeats a numerical sequence, or the decimal equivalent of fractions such as 2/3, which repeat endlessly, will do. The Universe Computer can't hold any one of them, because they go on indefinitely. We can out-imagine the universe's capacity to physically hold a complete representation of myriad, abstract thoughts that we casually deal with daily.

That is partly because we can also deal with another abstraction, limits. We can instruct our minds, or our computers, to limit the calculation of pi to, say, 4 digits, because after that it does not matter, leaving almost all of our memories to perform other tasks. Pragmatism rules! We can't perfectly solve the 3-body problem, such as landing a man on the moon. However, as Feynman said during a lecture at Hughes Research Labs, "Who cares? Just tell me how close you need to be and I will get you there." We have the capacity to think about *what it is we actually need*, rather than being defeated because we can't achieve perfect solutions to impossible problems.

Unlike DNA, which can only operate using its own concrete information, its human hosts can deal with both concrete and abstract information. We can imagine what we are not, or what we might want to be, and that is a phenomenal gift that is part and parcel of why are in a new stage of external information based, "enhanced "evolution.

What Has Enhanced Evolution Bought Us?

What do all of these differences between internal and external, abstract and concrete information mean? What do they buy us? Our species is only 200,000 years old. Our behavioral modernity, the point at which Homo sapiens began to demonstrate an ability to use complex symbolic thought, express cultural creativity and create written language, is only 50,000 years old. Look at what we have accomplished in that geologic instant of time, while we have remained basically the same genetically identical, physically undistinguished creatures that we were 50,000 years ago.

We are arguably, pound-for-pound, the weakest, slowest animal on the planet, and odds-on favorite for the least-likely-to-succeed, one-on-one, against anything. How do we puny ones stack up against our stronger, faster competitors after 50,000 years of our brain's superior capacity that gives us a comparative monopoly on utilizing external information to enhance our survival chances? It's pretty impressive.

We have used our externally stored information to create weapons to offset our physical disadvantages. We have used it to modify our environment by building things like shelters, power grids, and cars, which anyone can now build, anywhere. Ask Detroit. The principles, if not the actual blueprints, are all there for anyone to copy. If you don't have the know-how, YouTube will show you how to do anything from painting an oil portrait to building your own skating rink.

In our cars, any one of us can now outrun the fastest cheetah, and not just for 40 seconds, but for hours on end. Any one of us can use myriad pieces of construction equipment to outmuscle any elephant, and lift more and move more than any other animal could imagine, and do it relentlessly, 24 hours a day. No hummingbird can outrace or out-hover us, or falcon out dive us, or eagle outsoar us in our jet fighters, helicopters or commercial airliners.

By learning how to modify and/or recreate our environment, and take it with us, we can dive to the bottom of the Marianas Trench or perch atop Mt. Everest. If we had not learned how to cultivate and domesticate plants and animals for food and clothing, store and transport food and water for hundreds or thousands of miles, build fires and then erect shelters around them, none of us could survive a single Ohio winter as nature brought us into the world. Ditto for many of the other inhospitable places we now freely inhabit. We are even beginning to live for short times in space where almost no other creature can survive, let alone function.

With one exception, none of that is a triumph of DNA. The exception is that DNA gave us the cranial capacity to take our game to the next level. DNA gave us our consciousness along with a brain large enough to process, store, and retrieve complex information, which gave us a leg up on traditional evolution. If one wished to anthropomorphize, that might have been DNA's "goal" all along and it is quietly cheering us on. However, crediting DNA with a "purpose" is the same as injecting more unnecessary deities into the discussion, so I will take a pass on that creation myth. Again, what does it all mean?

It means that the ability to generate, manipulate and use EXTERNAL information is the equivalent of ENHANCED (stage 2) evolution.

Virtually all of our progress in out-surviving and out-performing our competitors has been accomplished by using external information to effectively accelerate, or enhance, our evolution. We are no longer on a traditional evolutionary track. We have elevated our game to a new level that traditional evolution, alone, cannot match. Our physical bodies are still the same wondrous, but limited, DNA built and driven machines. Meanwhile, our conscious processor, storing and retrieving information, at will, from a vast reservoir, external to our body, enables us to compete successfully with anything, anywhere on earth.

No animal can decide that it could use some wings and evolve itself into a bird. Every creature must wait until some fortuitous mutations enable it to fly so that it can inhabit that promising niche. We, on the other hand, can decide that we could use the *ability* to fly, and then invent winged vehicles to accomplish flight *for* us. We are independent of waiting for evolutionary happenstances to develop our capabilities. We can decide what we want to be *the equivalent of,* and *DO IT NOW!*

Our inventions are the necessary final step in the process of storing, disseminating, retrieving and *using* information to either gain advantage over our competitors or manipulate our environment to our advantage. If nobody does anything with all of this information, it is as useless as not having it. Inventions, from the wheel, to mathematics, to the printing press, to electric utilities and computers, are the executions of the process that give our external information advantage its power. They enable our accelerated evolution, because they enable us to use external information to outperform our competitors. Our inventions augment our capabilities and may increase our probability of extending our species survival. Without them we might have perished long ago.

When Does Enhanced Evolution Occur?

When does a life form transition from operating on purely internal information, or instinct, to when it begins to make choices based on external information? This is slippery to define, and is open to many variations of opinion, very similar to the discussion of when consciousness arises. Consciousness, alone, is insufficient to produce enhanced evolution.

A creature can be said to be conscious when it is self-aware; i.e., there is "me" and "non-me;" it can distinguish between "in here" and "out there." Simply being conscious, being aware that there is something else besides me, is insufficient to graduate to the domain of enhanced evolution. Membership implies that the organism makes choices to try to better enhance its odds for success. If external information is not utilized, enhancement does not occur, because other than being aware, nothing happens; nothing changes; therefore there is no evolution.

Memory is also required. Enhanced evolution occurs when a creature has not only become aware of the external, but has acquired and stored information from external sources in its memory, recalled it later on, and then *utilized* it to give itself a competitive advantage it did not already inherently possess. It requires a *learned, utilized* behavior, the benefit of which is to improve its odds for survival.

I have no idea what animal first demonstrated conscious, learned behavior and decision making, or when it occurred geologically. For now, it is obvious that humans are long past that genesis and have progressed almost geometrically since whenever it occurred. It has also occurred in many other life forms that react non-instinctively to their environment and display learned behavior. The difference between them and us is our superior brain capacity and organization that allows us to manipulate more complex data to more rapidly achieve more impressive, game-changing results.

How Far Have We Come?
Where Are We Going?

The information content of our external human libraries now includes the Library of Congress, the plans for every tool, weapon, vehicle, and structure we have ever designed, along with the methods for raising food and livestock developed over thousands of years. It includes mathematics, medicine, science, music, and literature along with the ability to analyze and compute unseen natural phenomena orders and orders of magnitude greater or smaller than our middling human scale. It is anything and everything we have ever learned, plus the ability to expand our external data banks to any size and transmit its benefits at the speed of light to every other member of our species for instant use, as needed.

We are no longer dependent on random, internal DNA mutations to enhance our well-being. To match the speed with which nature is going to challenge us to adapt we can utilize manipulation of our external information banks. We can achieve a level of enhanced evolution that can outperform DNA many times faster than its internal rate.

How rapid is it? Contrary to our intuition, which tends to perceive progress as linear, technological change has become exponential. This coincides with the cumulative, exponential growth of our external information banks. By one estimate, we won't experience 100 years of progress in the 21st century; it will be more like 20,000 years of progress.

Take just one indicator, Moore's Law, depicted in Fig. 2. Intel's Gordon Moore postulated that the rate of speed of computation would double every two years. Actually, the number of transistors available for calculations is the critical parameter, so microprocessor transistor count is charted in Fig. 2.

Figure 2, Microprocessor Transistor Counts

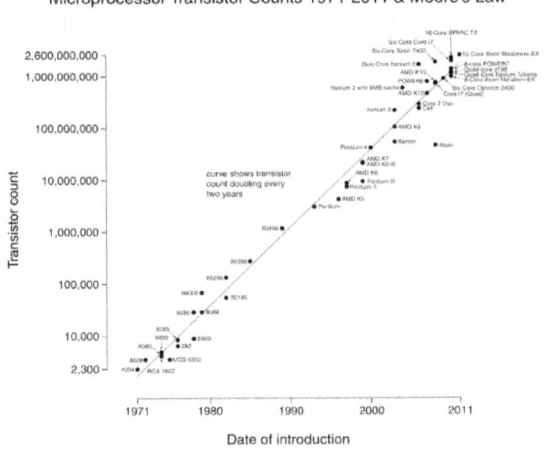

Microprocessor Transistor Counts 1971-2011 & Moore's Law

Fig. 2.
A plot of CPU transistor counts against dates of introduction; note the logarithmic vertical scale; the line corresponds to exponential growth with transistor count doubling every two years. From Wikipedia.

Was Moore right? Well, in the 40 years from 1971 to 2011 the number of transistors in a circuit doubled 20 times to a million times what it was. That's right on target with predictions. If you go back to 1900, calculation speed at the end of the 20th century was a trillion times faster than at its start. That's exponential progress!

A further implication of Moore's Law is that the cost of computation would drop about 20% every time the total number of circuits manufactured doubled. In addition to enormous capability increases, the cost keeps dropping at the same time. That principle is true for everything we mass-produce. Virtually any identical mass-produced item gets less expensive over time in real dollars, distributing its advantages to an ever-expanding user base by making it more affordable and available to ever more people.

And it is happening in every field. During this period of our history, our total knowledge accumulation is exponential, as is its distribution. We can now Google search millions of web sites per second to find answers to almost any question. The implications for human progress – what problems any one of us can undertake to solve, relatively quickly – are profound. DNA is no match for that.

Stage Three Evolution Is Already Here

We are at the point where we have so much information that we can even decode and begin to manipulate DNA, itself. I don't know if that is a good idea, or not, or what limitations we should, or should not, place on ourselves. The issues of how fast we can get ourselves into how much trouble, or wipe ourselves out, or what responsibilities we may have to each other, other life forms, and to our life-giving planet, merit our most thoughtful, separate discussions.

I postulate that if using external information is Enhanced, or Stage 2 Evolution, then gene manipulation, which we are already doing, is certainly Stage 3. That is definitely another game changer. When beings can decode themselves well enough to alter their own genes, then they can potentially modify their own physical structure, including appearance, resistance to disease, aging and brain capacity. That is essentially the ability to *reengineer existing creatures or create new creatures,* which qualifies as Stage 3 Evolution. We are taking unto ourselves what once would have been considered God-like capabilities. We will be tested to see if we have God-like judgment.

We are only at the threshold of Stage 3 evolution. We cannot begin to envision the consequences of that development, which are even more far-reaching than those of the Stage 2 evolution that enabled it. When human genetic engineering allows us to "breed" people of varying physical and mental characteristics and capabilities, we

47

may learn what Huxley knew; you can't run the world with all alphas. That realization implies that some form of Brave New World, with genetically engineered "classes" may not be far behind. Other than identifying that Stage 3 evolution is upon us, those are discussions for another day.

We are as Diego Rivera painted in his famous mural in Fig. 3, Man at the Crossroads. We have no idea what the limits of our external-information-based, accelerated evolution are, let alone those of Stage 3. One consequence

Figure. 3, Man At the Crossroads

Fig. 3 – Diego Rivera Fresco, Man at the Crossroads. It was commissioned to be in New York City's Rockefeller Center. The painting was controversial because it included an image of Lenin and a Soviet Russian May Day parade. Despite protests from artists, Nelson Rockefeller ordered its destruction before it was completed.

is that if used wisely those capabilities probably present our best potential to prolong our species lifetime into post-Earth, time periods. Again, we are a very fortunate life form, having evolved to live in our own "Goldilocks Haven" where everything is "just right" for our survival.

Realistically, that won't last forever, nor can our DNA evolve our way out of the vastly different environments and situations that are going to present themselves In Table 1. Our specie's Goldilocks zone has an expiration date. Over the very long term it seems that there is only one real option for prolonging our specie's survival – move on to other Goldilocks zones. DNA can't take us there. We will need a better tool, and we may already have it in hand.

In broad terms, we have the capacity to view Earth as basically our original, temporary spaceship. The odds are that we probably have a few million years to learn how to invent or recreate other spaceships that are:

- **Large enough**
- **With a sustainable**
- **Earth-substitute environment**
- **That lasts long enough**
- **And goes fast enough**
- **That we could migrate enough of us**
- **To another, younger, Goldilocks world**
- **Until it was time to move on again**
- **Which time will come no matter where we resettle.**

We might begin by occupying other solar system homes. Mars may become more earth-like as it heats up and with a little help from us Mars might buy us a few million years. Some moons further out, orbiting the gas giants may also become more habitable as the sun heats up. Eventually, we would have to learn enough to hopscotch our way around the galaxy to other planets in other solar systems to prolong our survival. That will be an extraordinarily formidable challenge, but we may have the capability to do just that.

That strategy only keeps working as long as we keep finding new planetary homes, because all planets eventually become inhospitable. The end game is to keep spreading us around so that a catastrophe on any one planet will not terminate the entire species.

Alas, eventually even that strategy will run its limits as the universe runs down and there is not enough energy density anywhere to support life. It reminds me of a version of the three laws of thermodynamics which, loosely stated, say that we are in a poker game where: 1) we can't win, 2) we can't break even, and 3) we're not allowed out of the game! In the meantime we may have as much as 40-50 million years (one-fourth of the time dinosaurs ruled) to solve our first major migration problem of leaving Earth. I'm optimistic that we can accomplish that, because that's 800 to 1,000 times longer than we have been able to use language to develop our current external information banks. As recently as 2,000 years ago, our ancestors could not even have fantasized the world we have created in so short a time so at our rate of knowledge growth there is hope that we can get the job done in the estimated time available.

Our best tool will be external, information-based, Second Stage Evolution. That elevates us to a new evolutionary capability level and will bail us out as best we can be bailed out. Stage Three Evolution may add to our potential for success, but we have no clue as to how that will play out.

So, the particular sequence of inventions in Table 2 is more than just inventions; it represents the ability to externally store, access and use information. It represents the equivalent of enhanced evolution, which can enable us to either out-survive all our competitors, or if misused, terminate us early. We surely have a tiger by the tail.

Acknowledgements

I must thank Jon Flinker for convincing me to prepare courses on The Big Bang and Inventions for The Baldwin-Wallace University Institute for Leaning in Retirement. Those preparations gave me the chance to crystalize my thoughts on evolution presented here. Several sources that also helped me to pull my thoughts together are: *The Evolution of Everything* by Matt Ridley, *The 100 Greatest Inventions of All Time* by Kendall Haven, *How Much Information Can Earth Hold*, Scientific American, August 2015, and *Why Information Grows*, both by Cesar Hidalgo. I also found Wikipedia to be very helpful on clarifying many points.

www.ingramcontent.com/pod-product-compliance
Lightning Source LLC
Chambersburg PA
CBHW071647170526
45166CB00003B/1463